Fract

By Robert W

An Easy Steps Math series book

Copyright © 2014 Robert Watchman

All rights reserved.

No portion of this publication may be reproduced, transmitted or broadcast in whole or in part or in any way without the written permission of the author.

Books in the Easy Steps Math series

Fractions
Decimals
Percentages
Ratios
Negative Numbers
Algebra
Master Collection 1 – Fractions, Decimals and Percentages
Master Collection 2 – Fractions, Decimals and Ratios
Master Collection 3 – Fractions, Percentages and Ratios
Master Collection 4 – Decimals, Percentages and Ratios

More to Follow

Contents

Introduction	7
Chapter 1 **Fraction Basics**	9
Chapter 2 **Types of Fractions**	19
Chapter 3 **Simplifying Fractions**	20
Chapter 4 **Adding Fractions**	27
Chapter 5 **Subtracting Fractions**	37
Chapter 6 **Multiplying Fractions**	44
Chapter 7 **Dividing Fractions**	50
Chapter 8 **Factions with Mixed Operations**	54
Chapter 9 **Fractions and Decimals**	61
Multiplication Tables	64
Answers	66
Glossary of Useful Terms	71

Introduction

This series of books has been written for the purpose of simplifying mathematical concepts that many students (and parents) find difficult. The explanations in many textbooks and on the Internet are often confusing and bogged down with terminology. This book has been written in a step-by-step 'verbal' style, meaning, the instructions are what would be said to students in class to explain the concepts in an easy to understand way.

Students are taught how to do their work in class, but when they get home, many do not necessarily recall how to answer the questions they learned about earlier that day. All they see are numbers in their books with no easy-to-follow explanation of what to do. This is a very common problem, especially when new concepts are being taught.

For over twenty years I have been writing math notes on the board for students to copy into a note book (separate from their work book), so when they go home they will still know how the questions are supposed to be answered. The excuse of not understanding or forgetting how to do the work is becoming a thing of the past. Many students have commented that when they read over these notes, either for completing homework or studying for a test or exam, they hear my voice going through the explanations again.

Once students start seeing success, they start to enjoy math rather than dread it. Students have found much success in using the notes from class to aid them in their study. In fact students from other classes have been seen using photocopies of the notes given in my classes. In one instance a parent found my math notes so easy to follow that he copied them to use in teaching his students in his school.

You will find this step-by-step method of learning easier to follow than traditional styles of explanation. With questions included throughout, you will gain practice along with a newfound understanding of how to complete your calculations. Answers are included at the end.

Chapter 1

Fraction Basics

Fractions are easy…
if you know how to work them out.

A fraction is simply a **part** of a **whole**. So if you have a whole pizza, one slice is just a part of the whole pizza. Two slices are also a part of the whole pizza, so are three slices, four slices, etc. eight slices becomes a whole pizza.

E.g.

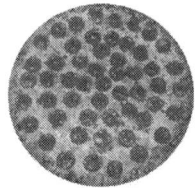  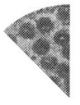

   1 whole pizza   1 whole pizza   one slice or
                            cut into 8 slices   one small part
                               or 8 parts   of the pizza

If you watch a movie, but turn the television off before the end, you've only watched a part of the whole movie. You might say to your friends that you "didn't watch the whole thing" and that you "only saw a part of it." This means you watched a fraction of the movie, or ate a fraction of the pizza. You may have watched most of the movie, or eaten most of the pizza, but it's still only a fraction.

The fraction can also be written as parts **out of** a whole. So, if the whole pizza is 8 pieces and you eat 3 pieces, then you've eaten three parts **out of** the eight, or three **out of** eight pieces, or three eighths $\frac{3}{8}$.

Another way to think of a fraction is as a **division**. But instead of writing the numbers on one line like $2 \div 3$ they are written one on top of the other like $\dfrac{2}{3}$.

So…

$2 \div 3$ means the same as $\dfrac{2}{3}$ and $4 \div 5$ means the same as $\dfrac{4}{5}$.

Each of these divisions can be written as a fraction

$$8 \div 3 = \dfrac{8}{3}$$

$$12 \div 5 = \dfrac{12}{5}$$

$$1 \div 2 = \dfrac{1}{2}$$

$$17 \div 9 = \dfrac{17}{9}$$

$$37 \div 49 = \dfrac{37}{49}$$

$$128 \div 105 = \dfrac{128}{105}$$

I think you get the idea.

Now you try some. Rewrite these divisions as fractions:

a) $6 \div 42 =$

b) $3 \div 21 =$

c) $2 \div 16 =$

d) $60 \div 12 =$

e) $6 \div 48 =$

f) $90 \div 9 =$

g) $7 \div 28 =$

h) $17 \div 45 =$

i) $9 \div 23 =$

j) $2 \div 8 =$

Now that you can turn a division into a fraction, you can do the opposite just as easily. All you have to do to change a fraction into a division is write the top number of the fraction first, then the division sign $(\div)$ then the bottom number of the fraction.

Here are some examples

$$\frac{1}{10} = 1 \div 10$$

$$\frac{4}{15} = 4 \div 15$$

$$\frac{7}{25} = 7 \div 25$$

Now you change these fractions to divisions:

a) $\dfrac{4}{9} =$

b) $\dfrac{11}{2} =$

c) $\dfrac{12}{2} =$

d) $\dfrac{12}{9} =$

e) $\dfrac{3}{8} =$

f) $\dfrac{3}{5} =$

g) $\dfrac{7}{3} =$

h) $\dfrac{10}{4} =$

i) $\dfrac{7}{49} =$

j) $\dfrac{8}{25} =$

Note: The top number of a fraction is called the NUMERATOR and the bottom number is called the DENOMINATOR.

$$\dfrac{3 \leftarrow \text{Numerator}}{4 \leftarrow \text{Denominator}}$$

**Check the glossary for other important words.

<u>Remember these two words. They will come up a lot.</u>

Learning multiplication tables helps with fractions because every fraction can be written in a number of different ways, that is, with different numbers, while still keeping its original value.

$$\frac{1}{2} = \frac{2}{4} = \frac{3}{6} = \frac{4}{8} = \frac{5}{10} = \frac{9}{18} = \frac{45}{90} = \frac{100}{200}$$

All these fractions above have the same value. They are all the same even though they have different numbers. They are called <u>equivalent fractions</u>.

Another way to explain this is that for each fraction the numerator and the denominator have been multiplied by the same number. Doing this doesn't change the <u>value</u> of the fraction, only the numbers.

For instance, starting with $\frac{1}{2}$ on the far left above, both the 1 (the numerator) and the 2 (the denominator) has been multiplied by 2 to get $\frac{2}{4}$.

And if both numbers are multiplied by 3 you will get $\frac{3}{6}$.

And if they have both been multiplied by 9 you will get $\frac{9}{18}$, and so on. But their values are still the same as the original fraction, which is $\frac{1}{2}$.

So, multiplying the numerator and the denominator by **the same number** makes equivalent fractions.

This is how you would do it:

$$\frac{2^{\times 2}}{5_{\times 2}} = \frac{4}{10}$$

$$\frac{4^{\times 3}}{7_{\times 3}} = \frac{12}{21}$$

$$\frac{3^{\times 7}}{11_{\times 7}} = \frac{21}{77}$$

$$\frac{5^{\times 4}}{9_{\times 4}} = \frac{20}{36}$$

$$\frac{6^{\times 5}}{7_{\times 5}} = \frac{30}{35}$$

$$\frac{8^{\times 9}}{13_{\times 9}} = \frac{72}{117}$$

Remember, the values of the fractions do not change when we multiply the numerator and denominator by the same number.

Now you change these fractions to their equivalent fractions:

a) $\dfrac{2 \times 3}{5 \times 3} =$

b) $\dfrac{6 \times 6}{5 \times 6} =$

c) $\dfrac{7 \times 4}{9 \times 4} =$

d) $\dfrac{3 \times 2}{3 \times 2} =$

e) $\dfrac{3 \times 2}{4 \times 2} =$

f) $\dfrac{5 \times 9}{6 \times 9} =$

g) $\dfrac{2 \times 7}{9 \times 7} =$

h) $\dfrac{2 \times 3}{7 \times 3} =$

i) $\dfrac{9 \times 5}{9 \times 5} =$

j) $\dfrac{9 \times 5}{10 \times 5} =$

Important Note:
It is important to remember that whatever we do to the numerator we must also do to the denominator and vice versa. That is, if we multiply the top by a number we must multiply the bottom by the exact same number. If we don't, then we change the value of the fraction and our answer will be wrong. We don't want this to happen.

Sometimes when doing calculations with fractions, you may get a question with a whole number like 3 or 7 or 19. Every number can be turned into a fraction by putting it over 1.

E.g. using the numbers above:

The whole number 3 can be turned into a fraction by putting it over 1.

Therefore $3 = \dfrac{3}{1}$

7 can be turned into a fraction by putting it over 1.

Therefore $7 = \dfrac{7}{1}$

And 19 can be turned into a fraction by putting it over 1.

Therefore $19 = \dfrac{19}{1}$

And the same can be done with any number.

So an expression like $\dfrac{4}{5} \times 100$ can become $\dfrac{4}{5} \times \dfrac{100}{1}$. You put the 100 over 1 and continue with the calculation.

Change these whole numbers into their fraction equivalents.

a) $37 =$

b) $245 =$

c) $54 =$

d) $592 =$

e) $93 =$

f) $2398 =$

g) $107 =$

h) $7632 =$

i) 456

j) a

Occasionally the numerator and denominator of a fraction will be the same number, this fraction is equal to 1. This applies to all numbers or anything else (see the last example below).

$$\frac{3}{3}=1, \quad \frac{11}{11}=1, \quad \frac{31}{31}=1, \quad \frac{257}{257}=1, \quad \frac{4678}{4678}=1, \quad \frac{x}{x}=1$$

(The last one is for when you learn algebra. The same rules still apply)

Therefore…

$\dfrac{2}{5}{\scriptstyle \times 3 \atop \times 3}$ is the same as $\dfrac{2}{5}\times\dfrac{3}{3}$ which equals $\dfrac{2}{5}\times 1$ which equals $\dfrac{2}{5}$

Whenever we multiply the numerator and denominator by the same number, to get an equivalent fraction, it just means that we are multiplying the fraction by 1. That is why the value of the fraction does not change.

Another way to show the above is to multiply it out, then simplify.

$$\frac{2}{5} \times \frac{3}{3} = \frac{2}{5} \times 1 = \frac{2}{5}$$

So anything you multiply by 1 stays exactly the same.

Work out (simplify) the following:

a) $\dfrac{121}{121} =$

b) $\dfrac{123}{123} =$

c) $\dfrac{913}{913} =$

d) $\dfrac{49}{49} =$

e) $\dfrac{241}{241} =$

f) $\dfrac{698}{698} =$

g) $\dfrac{27}{27} =$

h) $\dfrac{4574}{4574} =$

i) $\dfrac{5}{5} =$

j) $\dfrac{964}{964} =$

Chapter 2

Types of Fractions

There are three main types of fractions.

The first type is called a **proper fraction**. In a proper fraction the numerator is a <u>smaller</u> number than the denominator. Here are some examples of proper fractions.

$$\frac{2}{5}, \frac{4}{9}, \frac{12}{13}, \frac{3}{13}, \frac{2}{37}$$

The second type is called an **improper fraction**. In an improper fraction, the numerator is a <u>larger</u> number than the denominator. Here are some examples of improper fractions.

$$\frac{5}{3}, \frac{7}{5}, \frac{12}{4}, \frac{17}{3}, \frac{14}{13}$$

The third type is called a **mixed number**. A mixed number has a whole number and a proper fraction mixed together. You write the whole number first, then the fraction next to it. Here are some examples of mixed numbers.

$$3\frac{2}{7} \quad 2\frac{1}{2} \quad 4\frac{2}{5} \quad 5\frac{9}{17} \quad 37\frac{8}{19}$$

Whole Number   Proper Fraction

**When simplifying improper fractions, the answer is always a mixed number!

Chapter 3

Simplifying Fractions

When answering questions with fractions, it is best to write the answer in its simplest form. That means with the smallest numbers possible without changing the value.

Proper Fractions

To simplify a proper fraction, you would divide the numerator and the denominator by the largest number that fits into them both evenly. This is called the Greatest Common Divisor or GCD.

Here are some examples,

$$\frac{12^{\div 6}}{18_{\div 6}} = \frac{2}{3}, \quad \frac{21^{\div 7}}{77_{\div 7}} = \frac{3}{11}, \quad \frac{20^{\div 4}}{36_{\div 4}} = \frac{5}{9}$$

You can also ask yourself "what is the largest number that goes into the numerator and denominator without a remainder?" As you can see, for the above examples the answer is 6 for the first fraction, 7 for the second fraction and 4 for the third. The answers above are the simplest form of the fraction.

Remember that the simplified fraction is still an equivalent fraction, however it is written with smaller numbers.

Improper Fractions

An improper fraction can be simplified down to a mixed number by dividing the numerator by the denominator, or by following these steps.

E.g. Simplify $\dfrac{11}{3}$

Step 1. Ask yourself "how many times does the denominator go into the numerator"?

$11 \div 3 = 3$ r 2 Answer is 3 times, but there is a remainder of 2.

Step 2. Take the remainder 2 and put it over the denominator 3 to create a fraction.

I.e.

$\dfrac{2 \leftarrow \text{remainder}}{3 \leftarrow \text{Denominator}}$

Step 3. Now put the number 3 from step 1 and the fraction from step 2 together and you get $3\dfrac{2}{3}$. This is your final answer.

Here are some more examples.

a) Simplify $\dfrac{17}{4}$

Step 1. How many times does 4 go into 17?

$17 \div 4 = 4$ r 1 Answer is 4 remainder 1.

Step 2. Put the 1 over the 17 and you get the fraction $\dfrac{1}{17}$

Step 3. Put the number 4 from step 1 and the fraction from step 2 together to get $4\dfrac{1}{17}$. This is your final answer.

b) Simplify $\dfrac{25}{7}$

Step 1. How many times does 7 go into 25?

$25 \div 7 = 3 \text{ r } 4$ Answer is **3** remainder **4**.

Step 2. Put the 4 over the 7 and you get the fraction $\dfrac{4}{7}$.

Step 3. Put the number 3 from step 1 and the fraction from step 2 together to get $3\dfrac{4}{7}$. This is your final answer.

c) Simplify $\dfrac{54}{12}$

Step 1. How many times does 12 go into 54?

$54 \div 12 = 4 \text{ r } 6$ Answer is **4** remainder **6**.

Step 2. Put the 6 over the 12 and you get the fraction $\dfrac{6}{12}$.

Step 3. Put the number 4 from step 1 and the fraction from step 2 together to get $4\dfrac{6}{12}$.

Step 4. Since $\dfrac{6}{12}$ can be simplified down to $\dfrac{1}{2}$ your final answer is $4\dfrac{1}{2}$.

Mixed Numbers

A mixed number is **already simplified**, but it needs to be changed to an improper fraction in order to complete a calculation. You would do this by following these simple steps.

E.g. a) Change $3\frac{2}{3}$ to an improper fraction

Step 1. Multiply the whole number with the denominator

$3 \times 3 = 9$

Step 2. Add the answer from this to the numerator

$9 + 2 = 11$

Step 3. Put this answer over the original denominator from the question

$\frac{11}{3}$. Therefore $3\frac{2}{3} = \frac{11}{3}$ This is your final answer.

b) Change $5\frac{2}{7}$ to an improper fraction.

Step 1. Multiply the whole number with the denominator

$5 \times 7 = 35$

Step 2. Add the answer from this to the numerator $35 + 2 = 37$

Step 3. Put this over the original denominator from the question

$\dfrac{37}{7}$. Therefore $5\dfrac{2}{7} = \dfrac{37}{7}$ This is your final answer.

c) Change $3\dfrac{1}{2}$ to an improper fraction

Step 1. Multiply the whole number with the denominator

$3 \times 2 = 6$

Step 2. Add the answer from this to the numerator

$6 + 1 = 7$

Step 3. Put this over the original denominator from the question

$\dfrac{7}{2}$. Therefore $3\dfrac{1}{2} = \dfrac{7}{2}$ This is your final answer.

**To make fractions calculations easier, mixed numbers need to be changed to improper fractions first.

Now it's your turn to try some.

Simplify the following proper fractions

a) $\dfrac{16}{36} =$

b) $\dfrac{18}{27} =$

c) $\dfrac{25}{35} =$

d) $\dfrac{32}{48} =$

e) $\dfrac{9}{36} =$

f) $\dfrac{72}{81} =$

g) $\dfrac{37}{74} =$

h) $\dfrac{28}{112} =$

i) $\dfrac{44}{96} =$

j) $\dfrac{39}{52} =$

Simplify these improper fractions

a) $\dfrac{19}{5} =$

b) $\dfrac{61}{12} =$

c) $\dfrac{38}{9} =$

d) $\dfrac{11}{4} =$

e) $\dfrac{10}{3} =$

f) $\dfrac{39}{13} =$

g) $\dfrac{83}{12} =$

h) $\dfrac{117}{18} =$

i) $\dfrac{26}{11} =$

j) $\dfrac{325}{104} =$

Change these mixed numbers to improper fractions

a) $5\dfrac{2}{3} =$

b) $7\dfrac{2}{5} =$

c) $1\dfrac{1}{9} =$

d) $5\dfrac{1}{6} =$

e) $3\dfrac{3}{10} =$

f) $6\dfrac{3}{8} =$

g) $9\dfrac{4}{7} =$

h) $12\dfrac{2}{11} =$

i) $1\dfrac{3}{9} =$

j) $40\dfrac{7}{10} =$

Chapter 4

Adding Fractions

When you add fractions, you need to follow some basic steps.

The easiest fractions to add are the ones with the **same denominators**. All you need to do is add the numerators

E.g.

$$\frac{3}{13} + \frac{5}{13} = \frac{8}{13}$$

In the first example above, the denominators are both 13. So since they are the same, all we have to do is add the numerators. 3 plus 5 gives us 8, so our answer is $\frac{8}{13}$

$$\frac{12}{25} + \frac{5}{25} = \frac{17}{25}$$

In this second example, the denominators are both 25. Therefore we again add the numerators, 12 plus 5 gives us 17. So our answer is $\frac{17}{25}$.

Note that we **do not add the denominators.

Add these fractions that have the same denominator.

a) $\dfrac{2}{7} + \dfrac{3}{7} =$

f) $\dfrac{9}{43} + \dfrac{23}{43} =$

b) $\dfrac{23}{105} + \dfrac{39}{105} =$

g) $\dfrac{9}{25} + \dfrac{12}{25} =$

c) $\dfrac{23}{13} + \dfrac{7}{13} =$

h) $\dfrac{35}{61} + \dfrac{29}{61} =$

d) $\dfrac{25}{29} + \dfrac{8}{29} =$

i) $\dfrac{20}{37} + \dfrac{13}{37} =$

e) $\dfrac{7}{17} + \dfrac{9}{17} =$

j) $\dfrac{12}{27} + \dfrac{15}{27} =$

It does not matter if the fractions are proper or improper fractions, so long as the denominators are the same. In fact, the denominators **must** be the same so that we can add the numerators. Also, don't forget to simplify the answers if they are improper fractions.

Now that you know how to add fractions with the same denominator, the next step is to look at how to add fractions with **different denominators**. All the work you have done so far will help you with this.

If you have a question like:

$$\text{Calculate } \frac{1}{2} + \frac{1}{4} =$$

Step 1. Make the denominators the same. *(This is why you need to know the multiplication tables)* You do this by finding the best number that both denominators go in to evenly. In this case the best number is **4**. It is called the Least Common Multiple or LCM. For fractions it is also called the Least Common Denominator or LCD.

Step 2. Rewrite the question so that *both denominators are the same.* You do this by multiplying the top and bottom by the same number (to make equivalent fractions).

$$\frac{1}{2} + \frac{1}{4} = \frac{1^{\times 2}}{2_{\times 2}} + \frac{1^{\times 1}}{4_{\times 1}} = \frac{2}{4} + \frac{1}{4}$$

Step 3. Add the numerators

$$\frac{2}{4} + \frac{1}{4} = \frac{3}{4}$$

Your completed calculation should look like this

$$\frac{1}{2}+\frac{1}{4}=\frac{1^{\times 2}}{2_{\times 2}}+\frac{1^{\times 1}}{4_{\times 1}}=\frac{2}{4}+\frac{1}{4}=\frac{3}{4}$$

It is important to show your calculations or your working out because this shows the teacher that you understand the process and know what you are doing.

Another way is to multiply the denominators together.

So for the question, *calculate* $\frac{1}{2}+\frac{1}{4}=$

Step 1. You would multiply the denominators (2 x 4 = 8) and you would get

$$\frac{1}{2}+\frac{1}{4}=\frac{}{8}+\frac{}{8}=$$

Step 2. You need to work out the numerators. Whatever you did to each denominator to get 8, you **must** do to the numerator.

For the first fraction, the denominator is 2 and you multiplied it by 4 to get 8. So you must multiply this first numerator by 4 also (1 x 4 = 4) and put it over the first 8.

For the second fraction, the denominator is 4 and you multiplied it by 2 to get 8. So you must multiply this second numerator by 2 (1 x 2 = 2) and put it over the second 8.

You will end up with $\dfrac{1}{2} + \dfrac{1}{4} = \dfrac{4}{8} + \dfrac{2}{8} =$

Now you just <u>add the numerators</u> $\dfrac{4}{8} + \dfrac{2}{8} = \dfrac{6}{8}$.

Then simplify $\dfrac{6}{8} = \dfrac{3}{4}$, which is your final answer.

Here is another example.

Calculate $\dfrac{1}{2} + \dfrac{2}{5}$

Step 1. If the denominators are not the same, you must make them the same. To make them the same, you must find the best number that both denominators will go in to. This number is 10, (2 x 5 = 10). The 2 goes into the 10 five times and the 5 goes into the 10 two times. The question can be rewritten like so:

$$\frac{1}{2}+\frac{2}{5}=\frac{}{10}+\frac{}{10}$$

Step 2. For the first fraction, to find the new numerator, you must work out how many times the 2 goes in to the 10 (this is 5). Then multiply the numerator (1) by 5 and put the answer over the first 10.

$$\frac{1^{\times 5}}{2_{\times 5}}=\frac{5}{10} \quad \text{(Refer to equivalent fractions)}$$

The same is done with the second fraction, i.e. 5 goes into 10 two times, so the numerator is 2 x 2 = 4.

$$\frac{2^{\times 2}}{5_{\times 2}}=\frac{4}{10}$$

Now you have

$$\frac{1}{2}+\frac{2}{5}=\frac{5}{10}+\frac{4}{10}$$

Step 3. Since the denominators are now the same, all you need to do is add the numerators.

Therefore $\dfrac{1}{2}+\dfrac{2}{5} = \dfrac{5}{10}+\dfrac{4}{10} = \dfrac{9}{10}$

Remember

1. Make the denominators the same
2. Work out what the numerators will be
3. Add the numerators (Never the denominators)
4. Simplify if necessary

That is all there is to adding fractions.

There are times when a question will have more than two fractions to add and it could look something like this,

$$\dfrac{2}{3}+\dfrac{1}{4}+\dfrac{5}{6}$$

You would follow the exact same steps as above.

Step 1. Make the denominators the same. Find the lowest common divisor LCD. (In this case it is 12).

$$\dfrac{2}{3}+\dfrac{1}{4}+\dfrac{5}{6} = \dfrac{}{12}+\dfrac{}{12}+\dfrac{}{12} =$$

Step 2. Work out what the numerators will be. Whatever you did to the denominator to get 12, you must do to the numerator.

$$\frac{2}{3}+\frac{1}{4}+\frac{5}{6} = \frac{2^{\times 4}}{3_{\times 4}}+\frac{1^{\times 3}}{4_{\times 3}}+\frac{5^{\times 2}}{6_{\times 2}}$$

$$=\frac{8}{12}+\frac{3}{12}+\frac{10}{12}$$

Step 3. Add the numerators

$$=\frac{21}{12}$$

Step 4. Now simplify

$$=1\frac{9}{12}$$

Sometimes a question has mixed numbers such as $3\frac{1}{2} + 4\frac{2}{7}$,

The first thing you would do is change the mixed numbers to improper fractions, so:

$3\frac{1}{2} + 4\frac{2}{7} = \frac{7}{2} + \frac{30}{7}$, then follow the steps above

Your completed calculation would look like this with every step shown, including the simplification at the end.

$$3\frac{1}{2} + 4\frac{2}{7}$$

$$= \frac{7}{2} + \frac{30}{7}$$

$$= \frac{7^{\times 7}}{2_{\times 7}} + \frac{30^{\times 2}}{7_{\times 2}}$$

$$= \frac{49}{14} + \frac{60}{14}$$

35

$$= \frac{109}{14}$$

$$= 7\frac{11}{14}$$

Note how the calculations have been set out above. All the equals' signs have been written directly under one another. Try to follow this method. Your teacher or instructor will find it easier to read so there is less of a chance to lose marks.

Add these fractions with different denominators

a) $\dfrac{1}{3}+\dfrac{2}{5}=$

b) $\dfrac{2}{3}+\dfrac{1}{4}=$

c) $\dfrac{1}{4}+\dfrac{4}{5}=$

d) $\dfrac{1}{2}+\dfrac{2}{3}=$

e) $\dfrac{3}{4}+\dfrac{1}{3}=$

f) $\dfrac{1}{6}+\dfrac{2}{3}+\dfrac{1}{2}=$

g) $\dfrac{3}{8}+\dfrac{1}{3}+\dfrac{5}{6}=$

h) $2\dfrac{3}{7}+3\dfrac{2}{5}=$

i) $1\dfrac{4}{5}+2\dfrac{3}{8}=$

j) $3\dfrac{2}{7}+3\dfrac{11}{12}=$

Chapter 5

Subtracting Fractions

When you subtract one fraction from another, you follow the same basic steps as adding fractions.

If the denominators are already the same, you just subtract the numerators.

E.g.

$$\frac{8}{13} - \frac{5}{13} = \frac{3}{13}$$

As you can see above, where the denominators are both 13, you just subtract the 5 from the 8 to get the answer 3 and this number is written over the denominator 13. So your answer is $\frac{3}{13}$.

$$\frac{19}{25} - \frac{7}{25} = \frac{12}{25}$$

In the second example you would subtract the 7 from the 19 and you're left with 12, which goes over the 25. So your answer is $\frac{12}{25}$.

Subtract these fractions with the same denominators.

a) $\dfrac{3}{7} - \dfrac{2}{7} =$

b) $\dfrac{39}{105} - \dfrac{23}{105} =$

c) $\dfrac{23}{13} - \dfrac{7}{13} =$

d) $\dfrac{25}{29} - \dfrac{8}{29} =$

e) $\dfrac{9}{17} - \dfrac{7}{17} =$

f) $\dfrac{23}{43} - \dfrac{9}{43} =$

g) $\dfrac{12}{25} - \dfrac{9}{25} =$

h) $\dfrac{35}{61} - \dfrac{29}{61} =$

i) $\dfrac{20}{37} - \dfrac{13}{37} =$

j) $\dfrac{15}{27} - \dfrac{12}{27} =$

To subtract fractions with **different denominators**, you must first make the denominators the same just as you did for adding fractions.

So if you have a question like,

calculate $\dfrac{1}{2} - \dfrac{1}{4} =$

Step 1. Find the LCD.

Remember that you do this by finding the best number that both denominators go in to evenly. This is still called the least common denominator. In this case the best number is 4.

$$\frac{1}{2}-\frac{1}{4}=\frac{}{4}-\frac{}{4}$$

Step 2. Rewrite the question so that both denominators are the same.

Remember that you are using equivalent fractions to replace the fractions in the question.

$$\frac{1}{2}-\frac{1}{4}=\frac{1^{\times 2}}{2_{\times 2}}-\frac{1^{\times 1}}{4_{\times 1}}$$

Step 3. Subtract the numerators

$$=\frac{2}{4}-\frac{1}{4}=\frac{1}{4}$$

Another way is to multiply the denominators together.

For the question *calculate* $\dfrac{1}{2}-\dfrac{1}{4}=$

Step 1. Multiply the denominators (2 x 4 = 8) and you get

$$\frac{1}{2} - \frac{1}{4} = \frac{}{8} - \frac{}{8} =$$

Step 2. Work out the numerators. Whatever you did to the denominator, you **must** do to the numerator.

For the first fraction, the denominator is 2 and you multiplied it by 4 to get 8. So you multiply the first numerator by 4 also (1 x 4 = 8).

For the second fraction, the denominator is 4 and you multiplied it by 2 to get 8. So you multiply the second numerator by 2 also (1 x 2 = 2).

You end up with $\dfrac{1}{2} - \dfrac{1}{4} = \dfrac{4}{8} - \dfrac{2}{8}$

Now you just subtract the numerators

$$\frac{4}{8} - \frac{2}{8} = \frac{2}{8}$$

Finally you simplify $\dfrac{2}{8} = \dfrac{1}{4}$ This is your final answer.

Here is another example.

Calculate $\dfrac{2}{3} - \dfrac{3}{5}$

Step 1. Make the denominators the same. The best number that both denominators go in to (or the LCD) is 15. The 3 goes into 15 five times and the 5 goes into 15 three times.

We can rewrite the question like so:

$$\dfrac{2}{3} - \dfrac{3}{5} = \dfrac{}{15} - \dfrac{}{15}$$

Step 2. For the first fraction, to find the new numerator, you determine how many times the 3 goes into 15 (this is 5). You then multiply the numerator (2) by 5 and put the answer over the first 15.

$$\dfrac{2^{\times 5}}{3_{\times 5}} = \dfrac{10}{15}$$

The same is done with the second fraction, i.e. determine how many times 5 goes into 15 (this is 3), then multiply the numerator (3) by 3 and put the answer over the second 15.

$$\dfrac{3^{\times 3}}{5_{\times 3}} = \dfrac{9}{15}$$

So you have $\dfrac{2}{3} - \dfrac{3}{5} = \dfrac{10}{15} - \dfrac{9}{15}$ since the denominators are now the same, all you need to do is subtract the numerators.

Therefore $\dfrac{2}{3} - \dfrac{3}{5} = \dfrac{10}{15} - \dfrac{9}{15} = \dfrac{1}{15}$

**Subtracting three or more fractions and subtracting mixed numbers is done the exact same way as addition.

As you can see, subtraction is done the exact same way as addition, except there is a minus (-) sign instead of a plus (+) sign.

Subtract these fractions with different denominators

a) $\dfrac{2}{5} - \dfrac{1}{3} =$

b) $\dfrac{2}{3} - \dfrac{1}{4} =$

c) $\dfrac{4}{5} - \dfrac{1}{4} =$

d) $\dfrac{2}{3} - \dfrac{1}{2} =$

e) $\dfrac{3}{4} - \dfrac{1}{3} =$

f) $\dfrac{2}{3} - \dfrac{1}{6} - \dfrac{1}{4} =$

g) $\dfrac{5}{8} - \dfrac{1}{3} - \dfrac{1}{12} =$

i) $3\dfrac{4}{5} - 1\dfrac{3}{8} =$

h) $4\dfrac{3}{7} - 2\dfrac{2}{5} =$

j) $5\dfrac{2}{7} - 4\dfrac{4}{7} =$

Did you remember to simplify your answers for the questions above?

Chapter 6

Multiplying Fractions

Multiplying fractions is the easiest fraction calculation of all. All you need to do is:

1. multiply the numerators
2. multiply the denominators
3. simplify

E.g. Calculate $\dfrac{4}{5} \times \dfrac{3}{8}$

Step 1. Multiply the numerators (4 x 3 = 12)

$$\dfrac{4}{5} \times \dfrac{3}{8} = \dfrac{12}{}$$

Step 2. Multiply the denominators (5 x 8 = 40)

$$\dfrac{4}{5} \times \dfrac{3}{8} = \dfrac{12}{40}$$

Step 3. Simplify

$$\dfrac{12^{\div 4}}{40_{\div 4}} = \dfrac{3}{10}$$

**Remember that you must divide the numerator and the denominator by the <u>same number</u> when simplifying.

The final answer is $\dfrac{3}{10}$.

Special Note: The word 'of' in mathematics means times $(\times)$.

If a question asks, *calculate* $\dfrac{1}{3}$ *of* $\dfrac{4}{5}$ this is the same as $\dfrac{1}{3} \times \dfrac{4}{5}$

Answer these multiplication questions

Calculate

a) $\dfrac{1}{3} \times \dfrac{2}{5} =$

b) $\dfrac{2}{3} \times \dfrac{1}{4} =$

c) $\dfrac{1}{4} \times \dfrac{4}{5} =$

d) $\dfrac{1}{2} \times \dfrac{2}{3} =$

e) $\dfrac{2}{3} \times \dfrac{1}{6} \times \dfrac{1}{2} =$

f) $\dfrac{3}{7}$ *of* $\dfrac{2}{5} =$

g) $\dfrac{4}{5}$ *of* $\dfrac{3}{8} =$

h) $\dfrac{4}{3}$ *of* $\dfrac{2}{5} =$

i) $1\dfrac{1}{4} \times \dfrac{2}{3} =$

j) $\dfrac{4}{5} \times 1\dfrac{6}{7} =$

k) $1\frac{1}{2} \times 12 =$

n) $2\frac{3}{4} \times 6 =$

l) $2\frac{1}{4} \times 2\frac{1}{2} =$

o) $\frac{7}{8} \times 3\frac{1}{3} =$

m) $8 \times 3\frac{1}{4} =$

p) $6\frac{2}{5} \times 4\frac{1}{7} =$

If you have mixed numbers to multiply, change these to improper fractions first, and then multiply.

When multiplying fractions, you can sometimes simplify before you multiply. This is called <u>cross simplification,</u> where you can simplify the numerator of one fraction with the denominator of another fraction. You would do this to make calculations easier and faster, **but this can only be done with multiplication.**

E.g.

Calculate $\frac{1}{4} \times \frac{4}{5} =$

As you can see, there is a 4 in the first denominator and a 4 in the second numerator. These cancel each other out because they are the same as $\frac{4}{4}$ which equals 1.

So you would have $\frac{1}{4} \times \frac{4}{5} = \frac{1}{{}_1\cancel{4}} \times \frac{\cancel{4}^1}{5} = \frac{1}{5}$

If the numbers top and bottom are the same, you can just cancel them and replace them with 1s.

Here is a different example,

$$Calculate \ \frac{4}{7} \times \frac{1}{12} =$$

Working this out the normal way, you would:

1. Multiply the numerators,

$$\frac{4}{7} \times \frac{1}{12} = \frac{4}{}$$

2. Multiply the denominators,

$$\frac{4}{7} \times \frac{1}{12} = \frac{4}{84}$$

3. Simplify

$$\frac{4}{84} \begin{matrix} \div 4 \\ \div 4 \end{matrix} = \frac{1}{21}$$

If you <u>cross simplify</u> first, you would look for numbers, **top and bottom only,** that go into each other evenly.

So for our example, $\frac{\overset{1}{\cancel{4}}}{7} \times \frac{1}{\underset{3}{\cancel{12}}}$ you can see that 4 goes into 4 one time in the first fraction and 4 goes into 12 three times in the second fraction. Cross out the existing numbers and put in the new numbers to show this. You will be able to do this easily if you know your multiplication tables well.

The working out for the question becomes $\dfrac{1}{7} \times \dfrac{1}{3}$

Now multiply the numerators and the denominators.

$$\dfrac{1}{7} \times \dfrac{1}{3} = \dfrac{1}{21}.$$

The answer is the same as step 3 above. As you can see both methods are correct. Cross simplification just makes it a bit faster.

Mixed numbers should be changed to improper fractions, as this will also make calculations simpler.

E.g.

$$3\dfrac{1}{5} \times 3\dfrac{1}{8}$$

$$= \dfrac{16}{5} \times \dfrac{25}{8}$$

$$= \dfrac{\cancel{16}^{2}}{\cancel{5}_{1}} \times \dfrac{\cancel{25}^{5}}{\cancel{8}_{1}} = \dfrac{2}{1} \times \dfrac{5}{1} = \dfrac{10}{1} = 10$$

Use the cross simplification method to solve the following.

a) $\dfrac{5}{7} \times \dfrac{1}{10} =$

b) $\dfrac{3}{4} \times \dfrac{8}{11} =$

c) $\dfrac{4}{15} \times \dfrac{25}{16} =$

d) $\dfrac{25}{6} \times \dfrac{12}{5} =$

e) $\dfrac{27}{12} \times \dfrac{14}{21} =$

f) $\dfrac{30}{6} \times \dfrac{4}{20} =$

g) $\dfrac{8}{5} \times \dfrac{20}{12} =$

h) $\dfrac{33}{18} \times \dfrac{24}{44} =$

i) $\dfrac{6}{5} \times \dfrac{8}{3} =$

j) $\dfrac{8}{6} \times \dfrac{12}{4} =$

k) $3\dfrac{3}{4} \times 2\dfrac{2}{5} =$

l) $7\dfrac{1}{2} \times 4\dfrac{2}{5} =$

m) $\dfrac{18}{4} \times \dfrac{16}{9} =$

n) $\dfrac{8}{9} \times \dfrac{6}{20} =$

o) $\dfrac{3}{7} \, of \, \dfrac{2}{3} =$

p) $\dfrac{2}{3} \, of \, 24 =$

Don't forget to simplify your answers

Chapter 7

Dividing Fractions

Before dividing fractions, you must first learn about reciprocals. The **reciprocal** of a fraction is the fraction turned upside down.

The reciprocal of $\frac{3}{7}$ is $\frac{7}{3}$. The reciprocal of $\frac{1}{3}$ is $\frac{3}{1}$ which is 3.

The reciprocal of 8 is $\frac{1}{8}$ (remember 8 is the same as $\frac{8}{1}$).

The reciprocal of $3\frac{1}{3}$ is $\frac{3}{10}$. Change $3\frac{1}{3}$ to an improper fraction first then **invert** it (turn it upside down).

When dividing fractions such as $\frac{5}{7} \div \frac{1}{10} =$ all you do is change the division sign to a multiplication sign and write the reciprocal of the second fraction (turn the **second** fraction upside down).

E.g. a)

Calculate $\frac{5}{7} \div \frac{1}{10}$

Step 1. Change the division to a multiplication

$\frac{5}{7} \times$

Step 2. Write down the reciprocal of the second fraction (turn the second fraction upside down).

$$\frac{5}{7} \times \frac{10}{1}$$

Step 3. Multiply the fractions.

$$\frac{5}{7} \times \frac{10}{1} = \frac{50}{7}$$

Step 4. Simplify

$$\frac{50}{7} = 7\frac{1}{7}$$

Therefore

$$\frac{5}{7} \div \frac{1}{10} = \frac{5}{7} \times \frac{10}{1} = \frac{50}{7} = 7\frac{1}{7}$$

$7\frac{1}{7}$ is your final answer.

E.g. b)

Calculate $3\frac{1}{2} \div 1\frac{3}{4}$

Step 1. Change mixed numbers to improper fractions

$$\frac{7}{2} \div \frac{7}{4}$$

Step 2. Change division to multiplication and invert second fraction

$$\frac{7}{2} \times \frac{4}{7}$$

Step 3. Cross simplify then multiply

$$\frac{\cancel{7}^1}{\cancel{2}_1} \times \frac{\cancel{4}^2}{\cancel{7}_1} = \frac{2}{1} = 2$$

Find the reciprocal of the following and simplify where possible

a) $\dfrac{3}{4}$

b) $\dfrac{3}{15}$

c) $\dfrac{1}{17}$

d) 32

e) 15

f) $9\dfrac{3}{7}$

g) $5\dfrac{2}{3}$

h) $6\dfrac{2}{5}$

Calculate the following divisions. Remember to simplify your answer.

a) $\dfrac{1}{4} \div \dfrac{2}{3} =$

b) $\dfrac{8}{9} \div \dfrac{7}{6} =$

c) $\dfrac{8}{15} \div \dfrac{16}{15} =$

d) $\dfrac{5}{8} \div 20 =$

e) $\dfrac{4}{9} \div \dfrac{4}{9} =$

f) $6\dfrac{2}{7} \div 4\dfrac{1}{5} =$

g) $10\dfrac{1}{3} \div 12\dfrac{5}{7} =$

h) $\dfrac{3}{8} \div \dfrac{4}{5} \div \dfrac{2}{3} =$

i) $\dfrac{3}{7} \div \dfrac{6}{11} =$

j) $\dfrac{3}{4} \div \dfrac{9}{4} =$

k) $7 \div \dfrac{1}{4} =$

l) $4\dfrac{1}{5} \div 3\dfrac{3}{10} =$

m) $22 \div \dfrac{11}{15} =$

n) $\dfrac{3}{8} \div \dfrac{4}{5} \times \dfrac{2}{3} =$

Chapter 8

Fractions With Mixed Operations

In mathematics, there is a specific **order of operations**. This means that you cannot just add, subtract, multiply and divide numbers (or fractions) in the order that they are written. You must follow a specific order so that your answers will be correct.
The order is:

1. **Parentheses**: These can also be braces or brackets, "( ), { } or [ ]"
2. **Exponents**: These are squares and square roots, 3^2, $\sqrt{3}$
3. **Multiplication/Division**: 3 x 5 ÷ 7, these can be done together
4. **Addition/Subtraction**: 3 + 5 − 7 these can be done together

This becomes easier if you remember PEMDAS (or BODMAS in some countries).

Examples with whole numbers:

If you have a question like:

Solve $18 \div (8+1) \times 6$.

You must follow the specific order shown above.

First solve whatever is in the parentheses $(8+1)=9$.

This gives you

$18 \div (9) \times 6$ or $18 \div 9 \times 6$.

You can remove the parentheses once you solve whatever is inside of them. Since the only operations that are left are division and multiplication, these can be done together so you can solve these from left to right one step at a time.

The whole calculation would look like this:

Step 1. Complete the parentheses first

$$18 \div (8+1) \times 6$$

$$= 18 \div 9 \times 6$$

Step 2. Complete the division second (only because division and multiplication are what's left)

$$= 2 \times 6$$

Step 3. Complete the multiplication

$$= 12$$

If the question was *solve* $8 + (4+2) \times 7$,

Your working would be:

Step 1. Complete the parentheses first

$$8 + (4+2) \times 7$$

$$= 8 + 6 \times 7$$

Step 2. Complete the multiplication second (because the order of operations says multiplication before addition)

$= 8 + 42$

Step 3. Complete the addition

$= 50$

When working with fractions you follow the exact same order.

E.g. 1)

Calculate $3 + \dfrac{2}{5} \times \dfrac{1}{2}$

Step 1. Change whole numbers to fractions and change mixed numbers to improper fractions (this just makes working out easier. It is not part of PEMDAS).

$3 + \dfrac{2}{5} \times \dfrac{1}{2}$

Step 2. Complete the multiplication

$= \dfrac{3}{1} + \dfrac{2}{5} \times \dfrac{1}{2}$

$= \dfrac{3}{1} + \dfrac{2}{10}$

Step 3. Complete the addition Make equivalent fractions by making the denominators the same.

$$= \frac{3}{1} + \frac{2}{10}$$

$$= \frac{3^{\times 10}}{1_{\times 10}} + \frac{2}{10}$$

$$= \frac{30}{10} + \frac{2}{10}$$

Step 4. Add the numerators

$$= \frac{30}{10} + \frac{2}{10}$$

$$= \frac{32}{10}$$

Step 5. Simplify

$$= 3\frac{1}{5}$$

E.g. 2.

Calculate

$$\frac{5}{7} \times \left(\frac{2}{3} + 4 \div \frac{2}{3} \right) + 5 \times \frac{3}{7}$$

Although this looks complicated, it is done the same way using everything we have learned so far. Every step has been included. You probably won't need to do these many steps.

Step 1. Change whole numbers to fractions and change mixed numbers to improper fractions

$$\frac{5}{7} \times \left(\frac{2}{3} + 4 \div \frac{2}{3} \right) + 5 \times \frac{3}{7}$$

$$= \frac{5}{7} \times \left(\frac{2}{3} + \frac{4}{1} \div \frac{2}{3} \right) + \frac{5}{1} \times \frac{3}{7}$$

Step 2. Inside the parentheses, do the division first. (Remember to change the division sign to a multiplication sign and to invert or flip the second fraction).

$$= \frac{5}{7} \times \left(\frac{2}{3} + \frac{4}{1} \div \frac{2}{3} \right) + \frac{5}{1} \times \frac{3}{7}$$

$$= \frac{5}{7} \times \left(\frac{2}{3} + \frac{\overset{2}{\cancel{4}}}{1} \times \frac{3}{\cancel{2}_1} \right) + \frac{5}{1} \times \frac{3}{7}$$

$$= \frac{5}{7} \times \left(\frac{2}{3} + \frac{6}{1} \right) + \frac{5}{1} \times \frac{3}{7}$$

Step 3. Inside parentheses, now do the addition

$$= \frac{5}{7} \times \left(\frac{2}{3} + \frac{6}{1} \right) + \frac{5}{1} \times \frac{3}{7}$$

$$= \frac{5}{7} \times \left(\frac{2}{3} + \frac{6^{\times 3}}{1_{\times 3}} \right) + \frac{5}{1} \times \frac{3}{7}$$

$$= \frac{5}{7} \times \left(\frac{2}{3} + \frac{18}{3} \right) + \frac{5}{1} \times \frac{3}{7}$$

$$= \frac{5}{7} \times \left(\frac{20}{3} \right) + \frac{5}{1} \times \frac{3}{7}$$

Step 4. Remove parentheses and complete the multiplications

$$= \frac{5}{7} \times \frac{20}{3} + \frac{5}{1} \times \frac{3}{7}$$

$$= \frac{100}{21} + \frac{15}{7}$$

Step 5. To solve the addition make the denominators the same

$$= \frac{100}{21} + \frac{15}{7}$$

$$= \frac{100}{21} + \frac{15^{\times 3}}{7_{\times 3}}$$

$$= \frac{100}{21} + \frac{45}{21}$$

Step 6. Add the numerators

$$= \frac{100}{21} + \frac{45}{21}$$

$$= \frac{145}{21}$$

Step 7. Simplify

$$= 6\frac{19}{21}$$ This is your final answer.

Now you try the following fractions with mixed operations.

a) $2 + \dfrac{3}{4} \times \dfrac{1}{3} =$

b) $7 - \dfrac{3}{8} \div \dfrac{9}{10} =$

c) $2\dfrac{3}{4} - \left(4\dfrac{1}{2} - 3\dfrac{1}{4}\right) =$

d) $\dfrac{8}{9} \div \left(\dfrac{3}{4} - \dfrac{1}{8}\right) =$

e) $\dfrac{3}{4} \times \dfrac{2}{3} + \dfrac{5}{6} \times \dfrac{1}{2} =$

f) $\dfrac{3}{7} \times \left(\dfrac{1}{4} + 7 \div \dfrac{7}{9}\right) =$

g) $\left(5 \div \dfrac{5}{11} - \dfrac{7}{12}\right) \times \dfrac{1}{6} =$

h) $\dfrac{8}{9} \times \left(\dfrac{3}{4} + 2 \times \dfrac{1}{5}\right) + 2 \times \dfrac{1}{6} =$

Chapter 9

Fractions and Decimals

There are times when we need to **convert a fraction to a decimal** in order to complete certain calculations.
Remember at the start of the book it was said that a fraction is a division. So as an example, to convert $\frac{1}{5}$ to a decimal, just put $1 \div 5$ in your calculator and it will give you the answer 0.2. If you don't have a calculator, or you're not allowed to use one, you need to do the conversion using either short division or long division depending on the numbers.

To convert $\frac{1}{5}$ to a decimal without a calculator just follow these steps.

Step 1. You know that 5 doesn't go into 1, so you must put a decimal point and at least one zero after the 1 (see below).

$$5\overline{)1.0}$$

Step 2. Now start your division. Ask yourself "how many times does 5 go into 1"? The answer is zero, so you write a zero on the line directly over the 1 and a decimal point directly over the decimal point below (decimal points **must always** line up).

$$5\overline{)1.0}^{0.}$$

Step 3. Carry the 1 to the bottom zero to change it to a 10.

$$5\overline{)1.^10}^{0.}$$

Step 4. Now ask yourself "how many times does 5 go into 10"? The answer is 2. Write the 2 over the line next to the decimal point (above the 0).

$$5\overline{)1.^10}^{0.2}$$ The answer is 0.2

Now you try these:

Convert the following fractions to decimals using your calculator.

a) $\dfrac{1}{4} =$

b) $\dfrac{2}{5} =$

c) $\dfrac{3}{8} =$

d) $\dfrac{37}{100} =$

e) $\dfrac{12}{25} =$

f) $\dfrac{43}{50} =$

Convert these fractions to decimals <u>without</u> your calculator.

g) $\dfrac{1}{2} =$

j) $\dfrac{7}{35} =$

h) $\dfrac{3}{10} =$

k) $\dfrac{9}{20} =$

i) $\dfrac{5}{8} =$

l) $\dfrac{3}{24} =$

You can check your answers with a calculator when you finish, not before.

Multiplication Tables

To make calculations really easy, learn your multiplications tables. Here is a set of multiplication tables from 1 x 1 to 12 x 12 to help you if you need it.

1 x 1 = 1	2 x 1 = 2	3 x 1 = 3	4 x 1 = 4
1 x 2 = 2	2 x 2 = 4	3 x 2 = 6	4 x 2 = 8
1 x 3 = 3	2 x 3 = 6	3 x 3 = 9	4 x 3 = 12
1 x 4 = 4	2 x 4 = 8	3 x 4 = 12	4 x 4 = 16
1 x 5 = 5	2 x 5 = 10	3 x 5 = 15	4 x 5 = 20
1 x 6 = 6	2 x 6 = 12	3 x 6 = 18	4 x 6 = 24
1 x 7 = 7	2 x 7 = 14	3 x 7 = 21	4 x 7 = 28
1 x 8 = 8	2 x 8 = 16	3 x 8 = 24	4 x 8 = 32
1 x 9 = 9	2 x 9 = 18	3 x 9 = 27	4 x 9 = 36
1 x 10 = 10	2 x 10 = 20	3 x 10 = 30	4 x 10 = 40
1 x 11 = 11	2 x 11 = 22	3 x 11 = 33	4 x 11 = 44
1 x 12 = 12	2 x 12 = 24	3 x 12 = 36	4 x 12 = 48

5 x 1 = 5	6 x 1 = 6	7 x 1 = 7	8 x 1 = 8
5 x 2 = 10	6 x 2 = 12	7 x 2 = 14	8 x 2 = 16
5 x 3 = 15	6 x 3 = 18	7 x 3 = 21	8 x 3 = 24
5 x 4 = 20	6 x 4 = 24	7 x 4 = 28	8 x 4 = 32
5 x 5 = 25	6 x 5 = 30	7 x 5 = 35	8 x 5 = 40
5 x 6 = 30	6 x 6 = 36	7 x 6 = 42	8 x 6 = 48
5 x 7 = 35	6 x 7 = 42	7 x 7 = 49	8 x 7 = 56
5 x 8 = 40	6 x 8 = 48	7 x 8 = 56	8 x 8 = 64
5 x 9 = 45	6 x 9 = 54	7 x 9 = 63	8 x 9 = 72
5 x 10 = 50	6 x 10 = 60	7 x 10 = 70	8 x 10 = 80
5 x 11 = 55	6 x 11 = 66	7 x 11 = 77	8 x 11 = 88
5 x 12 = 60	6 x 12 = 72	7 x 12 = 84	8 x 12 = 96

9 x 1 = 9	10 x 1 = 10	11 x 1 = 11	12 x 1 = 12
9 x 2 = 18	10 x 2 = 20	11 x 2 = 22	12 x 2 = 24
9 x 3 = 27	10 x 3 = 30	11 x 3 = 33	12 x 3 = 36
9 x 4 = 35	10 x 4 = 40	11 x 4 = 44	12 x 4 = 48
9 x 5 = 45	10 x 5 = 50	11 x 5 = 55	12 x 5 = 60
9 x 6 = 54	10 x 6 = 60	11 x 6 = 66	12 x 6 = 72
9 x 7 = 63	10 x 7 = 70	11 x 7 = 77	12 x 7 = 84
9 x 8 = 72	10 x 8 = 80	11 x 8 = 88	12 x 8 = 96
9 x 9 = 81	10 x 9 = 90	11 x 9 = 99	12 x 9 = 108
9 x 10 = 90	10 x 10 = 100	11 x 10 =110	12 x 10 = 120
9 x 11 = 99	10 x 11 = 110	11 x 11 = 121	12 x 11 = 132
9 x 12 = 108	10 x 12 = 120	11 x 12 = 132	12 x 12 = 144

Answers

The following answers correspond to the sections above. Note that answers are written across the page and not downward like the questions.

Change Divisions to Fractions

a) $\dfrac{6}{42}$ b) $\dfrac{3}{21}$ c) $\dfrac{2}{16}$ d) $\dfrac{60}{12}$ e) $\dfrac{6}{48}$ f) $\dfrac{90}{9}$ g) $\dfrac{7}{28}$ h) $\dfrac{17}{45}$
i) $\dfrac{9}{23}$ j) $\dfrac{2}{8}$

Change Fractions to Divisions

a) $4 \div 9$ b) $11 \div 2$ c) $12 \div 2$ d) $12 \div 9$ e) $3 \div 8$
f) $3 \div 5$ g) $7 \div 3$ h) $10 \div 4$ i) $7 \div 49$ j) $8 \div 25$

Equivalent Fractions

a) $\dfrac{6}{15}$ b) $\dfrac{36}{30}$ c) $\dfrac{28}{36}$ d) $\dfrac{6}{6}$ e) $\dfrac{6}{8}$ f) $\dfrac{45}{54}$ g) $\dfrac{14}{63}$ h) $\dfrac{6}{21}$
i) $\dfrac{45}{45}$ j) $\dfrac{45}{50}$

Convert Whole Numbers to Fractions

a) $\dfrac{37}{1}$ b) $\dfrac{245}{1}$ c) $\dfrac{54}{1}$ d) $\dfrac{592}{1}$ e) $\dfrac{93}{1}$ f) $\dfrac{2398}{1}$ g) $\dfrac{107}{1}$
h) $\dfrac{7632}{1}$ i) $\dfrac{456}{1}$ j) $\dfrac{a}{1}$

Fractions with the Same Numerator and Denominator

a) 1 b) 1 c) 1 d) 1 e) 1 f) 1 g) 1 h) 1 i) 1 j) 1

Simplifying Proper Fractions

a) $\dfrac{4}{9}$ b) $\dfrac{2}{3}$ c) $\dfrac{5}{7}$ d) $\dfrac{2}{3}$ e) $\dfrac{1}{4}$ f) $\dfrac{8}{9}$ g) $\dfrac{1}{2}$ h) $\dfrac{1}{4}$ i) $\dfrac{11}{24}$ j) $\dfrac{3}{4}$

Simplifying Improper Fractions

a) $3\dfrac{4}{5}$ b) $5\dfrac{1}{12}$ c) $4\dfrac{2}{9}$ d) $2\dfrac{3}{4}$ e) $3\dfrac{1}{2}$ f) 3 g) $6\dfrac{11}{12}$ h) $6\dfrac{1}{2}$
i) $2\dfrac{4}{11}$ j) $3\dfrac{1}{8}$

Mixed Numbers to Improper Fractions

a) $\dfrac{17}{3}$ b) $\dfrac{37}{5}$ c) $\dfrac{10}{9}$ d) $\dfrac{31}{6}$ e) $\dfrac{33}{10}$ f) $\dfrac{51}{8}$ g) $\dfrac{67}{7}$ h) $\dfrac{123}{11}$
i) $\dfrac{12}{9}$ j) $\dfrac{407}{10}$

Adding Fractions with the Same Denominators

a) $\dfrac{5}{7}$ b) $\dfrac{62}{105}$ c) $\dfrac{30}{13}=2\dfrac{4}{13}$ d) $\dfrac{33}{29}=1\dfrac{4}{29}$ e) $\dfrac{16}{17}$ f) $\dfrac{32}{43}$
g) $\dfrac{21}{25}$ h) $\dfrac{64}{61}=1\dfrac{3}{61}$ i) $\dfrac{33}{37}$ j) $\dfrac{27}{27}=1$

Adding Fractions with Different Denominators

a) $\dfrac{11}{15}$ b) $\dfrac{11}{12}$ c) $1\dfrac{1}{20}$ d) $1\dfrac{1}{6}$ e) $1\dfrac{1}{12}$ f) $1\dfrac{1}{3}$ g) $1\dfrac{13}{24}$

h) $5\dfrac{29}{35}$ i) $4\dfrac{7}{40}$ j) $7\dfrac{17}{84}$

Subtracting Fractions with the Same Denominators

a) $\dfrac{1}{7}$ b) $\dfrac{16}{105}$ c) $\dfrac{16}{13}=1\dfrac{3}{13}$ d) $\dfrac{17}{29}$ e) $\dfrac{2}{17}$ f) $\dfrac{14}{43}$ g) $\dfrac{3}{25}$

h) $\dfrac{6}{61}$ i) $\dfrac{7}{37}$ j) $\dfrac{3}{27}=\dfrac{1}{9}$

Subtracting Fractions with Different Denominators

a) $\dfrac{1}{15}$ b) $\dfrac{5}{12}$ c) $\dfrac{11}{20}$ d) $\dfrac{1}{6}$ e) $\dfrac{5}{12}$ f) $\dfrac{1}{4}$ g) $\dfrac{5}{24}$

h) $2\dfrac{1}{35}$ i) $2\dfrac{17}{40}$ j) $\dfrac{5}{7}$

Multiplying Fractions

a) $\dfrac{2}{15}$ b) $\dfrac{1}{6}$ c) $\dfrac{1}{5}$ d) $\dfrac{1}{3}$ e) $\dfrac{1}{18}$ f) $\dfrac{6}{35}$ g) $\dfrac{3}{10}$ h) $\dfrac{8}{15}$

i) $\dfrac{5}{6}$ j) $1\dfrac{17}{35}$ k) 18 l) $5\dfrac{5}{8}$ m) 26 n) $16\dfrac{1}{2}$ o) $2\dfrac{11}{12}$

p) $26\dfrac{18}{35}$

Multiplying Fractions (cross simplification)

a) $\dfrac{1}{14}$ b) $\dfrac{6}{11}$ c) $\dfrac{5}{12}$ d) 10 e) $1\dfrac{1}{2}$ f) 1 g) $2\dfrac{2}{3}$ h) 1 i) $3\dfrac{1}{5}$

j) 4 k) 9 l) 33 m) 8 n) $\dfrac{4}{15}$ o) $\dfrac{2}{7}$ p) 16

Dividing Fractions (reciprocals)

a) $\dfrac{4}{3}=1\dfrac{1}{3}$ b) $\dfrac{15}{3}=5$ c) $\dfrac{17}{1}=17$ d) $\dfrac{1}{32}$ e) $\dfrac{1}{15}$ f) $\dfrac{7}{66}$

g) $\dfrac{3}{17}$ h) $\dfrac{5}{32}$

Dividing Fractions

a) $\dfrac{3}{8}$ b) $\dfrac{16}{21}$ c) $\dfrac{1}{2}$ d) $\dfrac{1}{32}$ e) 1 f) $1\dfrac{73}{147}$ g) $\dfrac{217}{267}$ h) $\dfrac{45}{64}$

i) $\dfrac{11}{14}$ j) $\dfrac{1}{3}$ k) 28 l) $1\dfrac{3}{11}$ m) 30 n) $\dfrac{5}{16}$

Mixed Operations

a) $2\dfrac{1}{4}$ b) $6\dfrac{7}{12}$ c) $1\dfrac{1}{2}$ d) $1\dfrac{19}{45}$ e) $\dfrac{11}{12}$ f) $3\dfrac{27}{28}$ g) $1\dfrac{53}{72}$

h) $1\dfrac{16}{45}$

Fractions and Decimals

a) 0.25 b) 0.4 c) 0.375 d) 0.37 e) 0.48 f) 0.86 g) 0.5 h) 0.3
i) 0.625 j) 0.2 k) 0.45 l) 0.125

Glossary of Useful Terms

Sum refers to addition. The sum of two numbers is the answer of one number **plus** another number. E.g. the sum of 2 and 6 is 8, (2 + 6 = 8).

Difference refers to subtraction. The difference between two numbers is the answer of one number **minus** another number. E.g. the difference between 6 and 2 is 4, (6 − 2 = 4).

Product refers to multiplication. The product of two numbers is the answer of one number **times** another number. E.g. the product of 2 and 6 is 12, (2 x 6 = 12).

Quotient refers to division. The quotient is the answer of one number being **divided** by another number. E.g. the quotient of 6 and 2 is 3 (6 ÷ 2 = 3).

Greatest Common Divisor (GCD) or **Greatest Common Factor** (GCF), or **Highest Common Factor** (HCF): is the largest number or factor that goes into two other larger numbers without remainders.

E.g.

Q. *What is the GCD of 18 and 24?*

The factors of 24 are 1, 2, 3, 4, 6, 8, 12 and 24.

The factors of 18 are 1, 2, 3, 6, 9, and 18.

The largest number that is in both sets above is 6. Therefore the greatest common divisor of 24 and 18 is 6.

Least Common Multiple or **Lowest Common Multiple** (LCM): is the smallest number or multiple that two other numbers can both go in to.

E.G.

Q. What is the LCM of 3 and 4?

The multiples of 3 are 3, 6, 9, 12, 18, 21, etc

The multiples of 4 are 4, 8, 12, 16, 20, etc

The lowest number that is in both sets above is 12. Therefore the lowest common multiple of 3 and 4 is 12.

Least Common Denominator or **Lowest Common Denominator** (LCD): is the smallest denominator that two other denominators can both go in to. *(Used when adding and subtracting fractions)*

E.g.

Q. What is the LCD of $\frac{1}{2}$ and $\frac{1}{3}$?

The multiples of the denominator 2 are 2, 4, 6, 8, 10, etc

The multiples of the denominator 3 are 3, 6, 9, 12, etc

The lowest number that is in both sets above is 6. Therefore 6 will be used as the lowest common denominator.

$$\frac{1}{2}+\frac{1}{3}=\frac{-}{6}+\frac{-}{6}$$

Note that the working out for LCM and LCD is exactly the same.

To get the **Reciprocal** of a fraction, just turn the fraction upside down. E.g.

If the fraction is $\frac{2}{3}$, then it's reciprocal is $\frac{3}{2}$.

The reciprocal of 3 is $\frac{1}{3}$. The reciprocal of $\frac{1}{3}$ is 3.

The get the reciprocal of a mixed number like $2\frac{1}{2}$, first change it to an improper fraction $\frac{5}{2}$, and then turn it upside down $\frac{2}{5}$. So the reciprocal of $2\frac{1}{2}$ is $\frac{2}{5}$.

Any fraction multiplied by it's reciprocal always equals 1.

i.e. $\frac{2}{5} \times \frac{5}{2} = 1$

In a division, there are three words that should be learned. These are:
The Dividend. the number being divided.
The Divisor. the number doing the dividing?
The Quotient. the answer.

This information can be shown with the division symbols.

$$Dividend \div Divisor = Quotient$$

or in a 'division box'

$$Divisor \overline{) Dividend }^{Quotient}$$

Or, as a fraction.
The <u>numerator is the dividend</u>, the <u>denominator is the divisor</u> and the <u>quotient is the answer</u>.

$$\frac{Dividend}{Divisor} = Quotient$$

Check out the other books in the series

Decimals
Percentages
Ratios
Negative Numbers
Algebra
Master Collection 1 – Fractions, Decimals and Percentages
Master Collection 2 – Fractions, Decimals and Ratios
Master Collection 3 – Fractions, Percentages and Ratios
Master Collection 4 – Decimals, Percentages and Ratios

These can all be found at Amazon.com

Printed in Great Britain
by Amazon